SCIENCE MILLIE
BUSY BEES!

by Brooklyn Collins

Sunshine & STEM
Read. Explore. Discover.

SCIENCE TIME!

It is Millie's favorite time!
Science time!
Today, Millie is going to learn about the busy life of honeybees with yoga.

First, Millie curls up like a tiny egg. Every honeybee starts as a tiny rice-sized egg in a hexagonal honeycomb cell.

Fun fact: The queen can lay more than 1,000 eggs in a day!

The egg hatched!
This only took 3 days.
Millie is now a baby bee, called a larva.
Millie grabs her legs,
a happy wiggly baby bee,
waiting for her first taste of royal jelly.

Millie is now a growing larva.
Nurse bees feed the larva bee bread,
a mixture of nectar and pollen.
Millie stretches her arms and legs,
growing fast.

Worker bees seal the cell with a wax cap.
Safe inside,
Millie tucks her arms and legs under her.
Now a pupa, she is
going through metamorphosis.

Metamorphosis means something amazing
is happening.
Tiny wings, legs, and antennae are growing.

Crunch! Crunch!
The young bee chews through the wax cap
and crawls out,
now an adult worker bee.
Millie stretches her arms up, reaching out of
the honeycomb cell.

Millie stands up tall.
Millie is an adult bee now and ready to work.
Worker bees have many important jobs.
As they grow older, their jobs change.

One of the worker bees' first jobs is cleaning.
Millie helps keep the hive neat and healthy.

Next, Millie becomes a nurse bee.
She helps feed and care for baby bees.

Fun fact: Nurse bees make special food called royal jelly. Some baby bees are only given royal jelly, which tells the baby bee's body to become a queen bee.

Now Millie helps build a honeycomb.
Worker bees make wax from special glands
on their bodies.
Millie stretches out her arms as if she were
building each cell into a hexagon!

Millie stands watch at the hive entrance.
Guard bees help protect the colony.

Fun fact: Guard bees use their sense of smell to recognize hive members.

Now this bee gets to go out on an adventure.
Millie stretches her arms to fly from flower to flower, collecting nectar and pollen.

Fun fact: Honeybees can visit hundreds of flowers in one day.

Buzz! Wiggle! Waggle!
When Millie finds flowers, she performs a waggle dance.
The dance tells other bees where to fly.

Back at the hive, Millie stretches to her toes to place nectar into a honeycomb cell.

Finally, Millie fans her wings to dry the honey.
The honey is now stored for the whole hive to enjoy.

Every bee has an important job.
Together they build, protect, and care for their hive.

For my little scientists,
Whose joy and curiosity
inspire me every day.

Written by Brooklyn Collins
Published by Sunshine & STEM
ISBN: 979-8-951584-01-4
sunshineandstem.com

www.ingramcontent.com/pod-product-compliance
Lightning Source LLC
LaVergne TN
LVHW070937210926
848482LV00003B/266
* 9 7 9 8 9 5 1 5 8 4 0 1 4 *